an
Earthworm
is born

By William White, Jr., Ph.D.
PHOTOGRAPHS BY THE AUTHOR

**STERLING
NATURE
SERIES**

STERLING PUBLISHING CO., INC.   NEW YORK

Oak Tree Press Co., Ltd. London & Sydney

STERLING NATURE SERIES

AN ANT IS BORN

A BEE IS BORN

A BIRD IS BORN

BIRDS THAT FLY IN THE NIGHT

A BUTTERFLY IS BORN

AN EARTHWORM IS BORN

A FERN IS BORN

A FROG IS BORN

A FRUIT IS BORN

THE HIDDEN LIFE OF FLOWERS

THE SECRET LIFE OF SMALL ANIMALS

A SILKWORM IS BORN

TINY LIVING THINGS

A TREE IS BORN

A TREE GROWS UP

A TURTLE IS BORN

Acknowledgments

The author and publisher wish to thank William White III and James M. White for their unstinting toil in collecting, counting and caring for the myriad earthworms which were studied in this book; Sara A. White for the fine detailed dissections and Rebecca L. White for preparing the manuscript; and the officials of Wild-Heerbrugg Ltd., Heerbrugg, Switzerland, for the use of one of their Model M-20 research microscopes.

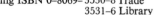

CONTENTS

Illus. 1. An earthworm makes its way through its burrows under an old rotted log. The burrows are moist and slippery with a thin film of water.

INTRODUCTION

We have all seen earthworms. They are among the first animals that small children touch and play with by themselves. Found almost anyplace where soil and green plants exist, they live in gardens and pastures as well as in parks and grassy plots in cities. They can be found almost everywhere in the ·world except in deserts where it is too dry, and in the polar regions where it is too cold. Earthworms seem strange because they do not have the parts or shapes we expect animals to have. They have no eyes, ears, or nose, no bones or legs and they move in a very strange way.

Earthworms spend all day under the ground in burrows, and come to the surface at night or when forced out of their burrows by too much rain. During their nightly travels on the surface they take small parts of rotten leaves and other foodstuffs down into their burrows. Also they come out at night to look for mates living in the burrows near their own (Illus. 1).

Earthworms are among the most important animals on earth for the survival of both plants and animals, since they act as reducers. They reduce dead matter to a form that can become living material again. This means that they break down dead plant materials and turn them into usable humus by eating them and passing them through their bodies. Earthworms also serve as one of the most important links in the food chain—they feed on microscopic matter and in turn are eaten by larger species. The earthworm is one of the most common foods for birds. However, many other animals are also dependent on the worm supply for food. Although earthworms do not seem to be very interesting at first, if you look closely and watch them you will soon discover that they have a fascinating life cycle all their own.

Earthworms have interested scientists for many hundreds of years. The British biologist, Charles Darwin, studied them for many years and wrote a whole book about them. In 1881, he counted the earthworms in a British pasture and found that there were about 23,000 per acre. He pointed out that the ploughing action of earthworms was very important. By taking plant material from the surface down to lower levels of the soil and by bringing up digested material and minerals from below, the earthworm constantly fertilizes the soil.

While everyone has seen earthworms, few people have taken the time to find out very much about them. Their egg cases are very common and can be found within a few inches of the surface of any grassy plot. Their burrows are one of the most common sights in any shallow hole in the ground. As you read about worms, go out in the garden or park and look for some. In the back of this book are some basic experiments with worms that you can try for yourself!

LIFE CYCLE OF THE EARTHWORM

Habitat

In the spring, in temperate climates, earthworms come up from the deeper burrows where they have spent the winter. They first begin to appear in early spring and may be found just below the surface up until mid-autumn. At night they come to the surface and excrete their waste products in forms called "casts." When they surface, they usually keep their back or posterior end firmly anchored in their burrows and only move about their small territory with their front or anterior end. At the slightest sign of light or a vibration in the ground, they withdraw their front ends back into their burrows. However, they may be found all the way out of their burrows on hot, humid nights (Illus. 2).

The earthworm is nocturnal, a night animal, and it has very sensitive skin, that cannot stand the direct rays of the sun. In fact, the animal breathes directly through the skin and so must stay moist and out of direct heat or light. Even though worms can survive under water for some time they are often washed out of their burrows by heavy rain. It is then that they can be found in great numbers in streets and other paved surfaces.

Illus. 2. A photographic light catches earthworms of all sizes out on the surface of the ground at night searching for food.

When a great many earthworms are forced out of their burrows, the birds have a feast. Many other animals besides birds look for earthworms on rainy days. Turtles (Illus. 3), snakes, frogs and toads wander far from their home areas on wet days to hunt earthworms. For protection, an earthworm cannot stay on the surface, but must soon find another burrow that is dry or another shady hiding place, or it will die. The ultra-violet rays of the sun are deadly to earthworms and many other soil-dwelling creatures. The degree to which earthworms are sensitive to light and to what wavelength of light is not fully understood. It is obvious from the speed with which they jerk back into their burrows, that they can detect light very quickly. They are definitely less sensitive to dark red or blue light than to white light.

Earthworms are very sensitive to sound vibrations in the soil. The presence of a garden snake (Illus. 4) or mole, both of which eat a large number of earthworms, is sensed by the earthworms which may come to the surface to escape. It has been known that many kinds of man-made noise, especially low-pitched noise will draw earthworms out of their burrows. In certain

Illus. 3. A box turtle turns over wood chips hunting for earthworms.

Illus. 4. A garden snake finishing a meal of a large earthworm.

parts of the southern United States where many earthworms are sold for fishing bait, there are people

Illus. 5. When disturbed, earthworms often move out at night across the surface of the ground, to other burrows nearby.

who know just how to drive heavy wooden stakes into the ground and "play them" with other pieces of wood to make a particular sound. This sound either draws or forces the earthworms to the surface where they are collected for sale (Illus. 5).

Earthworms are found in moist soils in all parts of the world, but they prefer packed soil in an area which has not been dug up or disturbed. They will surface under rocks, logs, or rotting piles of plants where they can get to the air and yet stay hidden from the sun.

Illus. 6. A flagstone piece has been removed from its location on bare soil to reveal a small earthworm pulling back into its burrow to avoid the light.

Illus. 7. The nearly hidden entrance to an earthworm's burrow in old settled forest soil. A dry, partially consumed leaf is at the entrance, which will be broken up and taken underground at nightfall.

Turning over rocks and small pieces of wood in the forest or garden will often reveal many earthworms of various sizes. It is possible to collect worms that come to the surface under one large stone on a regular basis (Illus. 6). If an earthworm is removed from a burrow at the soil surface, it takes only a day or so until its place is taken by another one.

Most animals, whether they are insects, reptiles, birds, or mammals, set out ''territories,'' small areas for themselves. They do not let any other individual of their species into the area except during breeding season. This does not seem to be the case with the earthworm. It appears that the new generations simply spread out to newer areas, or areas that have lost their earthworm population through flooding or the action or worm-eating predators.

Careful study shows that earthworms tend to travel through the same burrows and will feed on material they find on the surface as long as food can be found. When there is not enough food, the earthworm will bulk feed. Bulk feeding is the process of taking in earth through the mouth, or prostomium, and taking out of it tiny bits of animal and vegetable material, the organic portions. After the useful portions are eaten, the sand grains and gritty clay are passed on through the digestive system and out through the anus into

Illus. 8. A close-up of an earthworm burrow after the dead-leaf cover and small pebbles have been removed. The half-burrow, or trench, leading to the entrance is usually hidden.

Illus. 9. The posterior end of a very large earthworm being poked out of a burrow to excrete a casting. This activity is taking place at the corner of an observation frame.

the soil (Illus. 9 and 10). The earthworm consumes vast quantities of decaying leaves, algae and tiny soil animals in this way (Illus. 11).

By burrowing deep and curling into a tight circle, earthworms can easily live over the winter below the frost line in most of their range. The frost line may vary from 10 cm (4 inches) in warm climates to over 1.5 m (58 inches) in colder areas. However, it is not so much the cold and winter freezing which limit the earthworm's range as the danger of dryness. In semi-desert areas the earthworm may have to burrow down

Illus. 10. An ultra-close-up of a freshly deposited casting.

into the soil as much as 2.1 m (7 feet) to keep from drying out. In such areas the earthworm does not have to hibernate (sleep through the winter) but it must estivate (pass the summer in a dormant state).

While some species of earthworm build protective piles of pebbles around the entrance to their burrows, most worms simply attempt to come up under some object such as a rock or a board. Protected by these coverings, the worms will sometimes feed on the surface. They sweep the mouth or *prostomium* across the surface of the earth in a fashion very similar to a vacuum cleaner. Charles Darwin noted in his book entitled,

The Formation of Vegetable Mould Through the Action of Worms, that they would take many types of materials down in their burrows, pointed end first. It has been shown that pine needles and even tiny bits of paper are handled the same way.

Reproduction

The worm's method of reproduction works very well, so well that it is not unusual to find plots of ground where there are several hundred thousand

Illus. 11. A high-speed night photograph of an earthworm caught in the act of consuming a dried leaf on the surface of the ground.

Illus. 12. Some of the fungi and partially decayed leaves around the outer entrance to an earthworm burrow.

earthworms, some of which may have lived for as long as 10 years. However, it is not likely that many live that long because the earthworm has a great many natural enemies.

The earthworm hatches from an egg which is laid in the soil, along with many other eggs, in an egg case. Although these egg cases are very common, most

Illus. 13. An earthworm egg case nearing hatching, the ends are plugs of hardened slime. (10x)

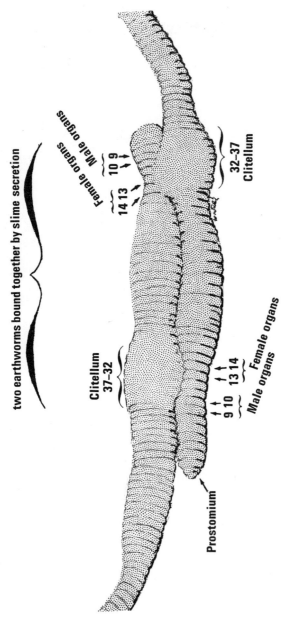

Illus. 14. Simultaneous mating of two earthworms bound together by slime secretion. The numbers indicate the rings encircling the various organs.

Illus. 15. Two mature worms mating. The seminal openings are enclosed in the partner's thickened clitellum. The male organs of the lower earthworm are at the left.

people do not recognize them when they see them. The cases are about the size and shape of a small pea, usually yellow-brown and hard and rubbery to the touch (Illus. 13). Sometimes the egg cases of earthworms are called cocoons, but this is not correct. The newly hatched earthworms are identical in appearance to their parents, but not much thicker than a piece of string. Of a pink to red shade, they are lighter in coloration than are mature worms. It takes them nearly a year to reach maturity, but they begin burrowing into new areas almost as soon as they hatch.

Each worm has both male and female sexual organs. The male sex cells or sperms are ready before the

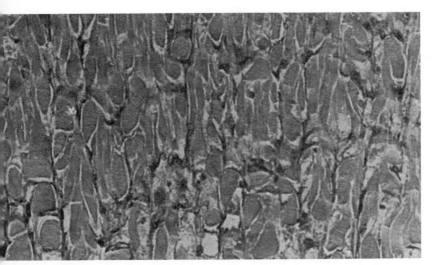
Illus. 16. The slime-filled cells and spaces of the swollen clitellum.

female sex cells or eggs. The openings or ducts through which the sperms are exchanged with other earthworms are positioned so that no worm can fertilize itself. Two worms mate by coming together in a side-by-side and head-to-tail position. Often both worms still have their posterior or back ends anchored into their burrows, if the burrows are near enough to each other (Illus. 14 and 15).

The two earthworms then form a slime coating around themselves which serves to bind them together. Each worm contracts to squeeze out small spurts of sperm cells. These travel down the sides of the worm in small channels between the worm's body and the slime coating. The sperm cells enter the ducts of the other worm, and are held until the eggs are ready to be laid. The two worms break away after they have exchanged the male sex cells or sperm. Some time later the eggs are ready to be laid. The worm forms a thick ring of slime around its body with a special organ called the *clitellum* (Illus. 16). Into this ring the earthworm squeezes out its own eggs and the sperm it has received from the other earthworm in mating. Finally the worm simply moves out of the ring, leaving it in the soil with the fertilized eggs developing inside. The ends close and the slime hardens into the egg case.

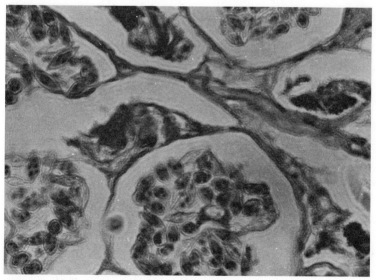

Illus. 17. Male sperm cells developing in the testes.

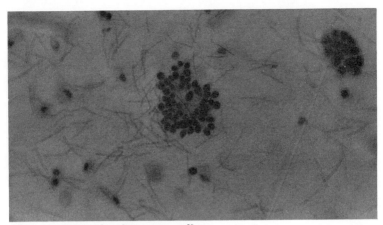

Illus. 18. Free-swimming sperm cells.
Illus. 19. The oviduct through which the eggs are ejected into the egg case with the sperm.

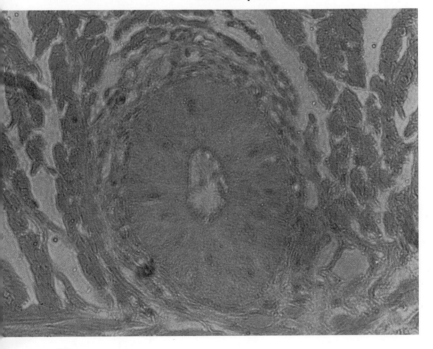

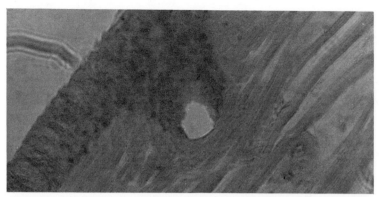

Illus. 20. Opening of the seminal vesicle passing through the epidermis.

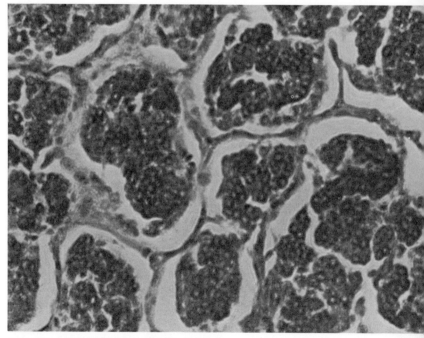

Illus. 21. Eggs developing within the ovary.

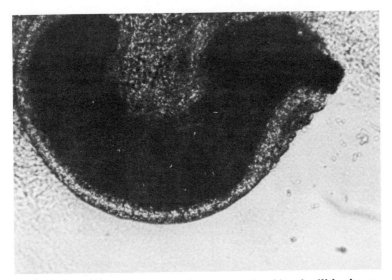

Illus. 22. An earthworm embryo only a few weeks old and still in the egg case. The segments are just visible on the edges of the outer layers of cells through the transparent epidermis. (100x)

The egg case contains a milk-white sticky substance which both protects and nourishes the embryos. While there are always several eggs and more than enough sperm within each egg case, only one or two worms usually hatch. The time it takes for the embryo to develop seems to depend both on the temperature and the amount of moisture in the soil (Illus. 22). The egg case shrinks slightly before hatching and turns a darker yellow-red before the earthworms hatch out. The total length of time from egg laying until hatching may run from 30 to 100 days.

The lighter color of the newly hatched worm (Illus. 23) is explained by the fact that it has not

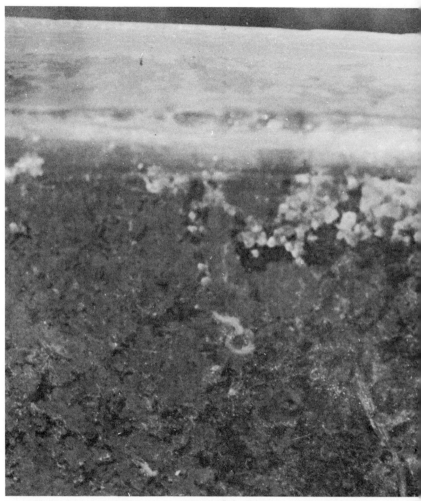

Illus. 23. A newly hatched pinkish earthworm (6-shaped figure in middle of photograph). The hatchling is making its way upward through the soil in an observation frame.

eaten any natural material and gained any coloring matter from the waste products of digestion. The

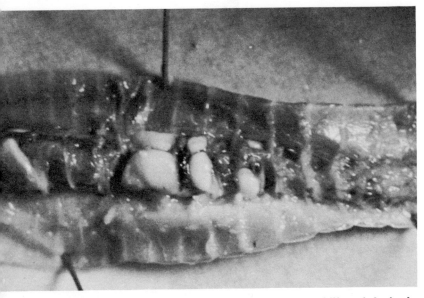

Illus. 24. A dissection of a large earthworm from the midline of the back.

newly hatched worm is about 12mm ($\frac{1}{2}$ inch) in length and about 5mm ($\frac{1}{32}$ inch) in diameter. If you examine an earthworm, you will notice that its body is composed of many ring-like segments. The baby earthworms vary in the number of rings or *annula*, but most seem to have between 89 and 114.

Life Span

The newly hatched earthworm must grow for nearly a year before it is mature enough to lay eggs of its own. Earthworms have been kept in captivity for nearly 7 years and some species are known to live 10 years.

There are many different species of earthworm, of which the most common is called *Lumbricus terrestris* by scientists. There are several phyla that include animals popularly called worms, but all earthworms are grouped in the phylum *Annelida*, or worms with rings around their bodies. The rings have nothing to do with the age of the worm. Earthworms do not grow more rings as they get older. Earthworms cannot regenerate new parts, except for a few *annula* or segments, at the front, or anterior end and perhaps 3 or 4 at the back, or posterior end (Illus. 25). If earthworms are cut in half or injured anywhere from the 4th to the 23rd segment they will die.

Illus. 25. A very large, mature earthworm. Note the large, stretched segments.

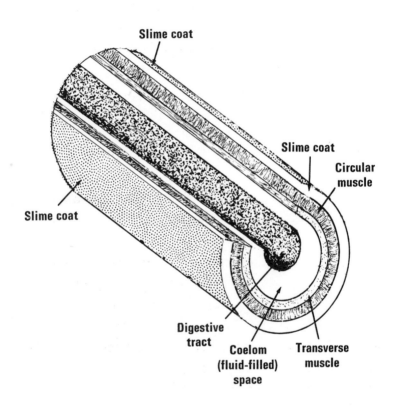

Slime coat

Slime coat

Circular muscle

Slime coat

Digestive tract

Coelom (fluid-filled) space

Transverse muscle

Illus. 26. Schematic drawing of earthworm's tubes within tubes.

ANATOMY OF THE EARTHWORM

The earthworm is classified by science as an invertebrate, since it has no backbone. Like other annelids, or segmented worms, earthworms have a soft, pliant outer layer of slime which is formed by special cells in the outermost skin or *epidermis*. As we have noted the earthworm has a specific front end or *anterior* and back end or *posterior*. It is also bilaterally symmetrical, that is, it has the same organs in pairs on both sides of its body.

The best way to understand how the earthworm lives and moves is to think of 5 tubes, one inside the other and all of about the same length (Illus. 26). The outermost tube is made up of a slimy secretion and serves three functions. It keeps the inner tubes moist and protects them from drying out. It acts to make the surrounding surface slippery so that the other tubes can slide through it more easily. It protects the inner tubes from infection by micro-organisms. Since there are many millions of these in the soil, infection from fungi or bacteria is one of the greatest dangers to earthworms.

The second tube contains the rings of muscle which go around the worm, and are the segments which you can see. They range in number from 89 up to 156 in individual earthworms, but the average is 114. Each ring is separated from the next one by a wall of muscle cells which helps to keep any movement contained within the ring. These are called *septa* or segment walls.

The next tube inside the circular rings of muscles, the septa, are the long muscle fibres which run from the head to the tail. These are called the *transverse muscles* and they pull or push in the same way from the head to tail. They can shorten any segment or relax and be pushed out by the constriction of the circular muscles.

The fourth tube is a long cigar-shaped space filled with fluid, called the *coelom*. However, because liquids cannot be squeezed into smaller spaces, the coelom is really rather solid and unyielding. The muscles of the two outer tubes squeeze or pull against it. The result is that the outer circular rings of muscle squeeze the transverse tube of muscle against the coelom and cause it to move forward (Illus. 27). The action is similar to squeezing toothpaste out of a tube. The transverse muscles constrict and pull the back end of the worm forward toward the head. Small hard, horny projections, something like claws are pushed out from the two muscular tubes to anchor the forward part of the worm while its back is pulled up. In this fashion the earthworm literally pumps itself through the soil. The small horny projections are called *setae* and are also

Illus. 27. An earthworm moves over the unfamiliar surface of a sheet of glass by pushing its prostomium out as far as it can reach, to investigate the area.

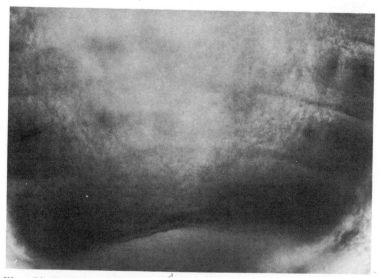

Illus. 28. A microscopic view of the 1st through the 4th segments of the nearly developed earthworm embryo. The setae are just beginning to become visible.

Illus. 29. The mid-section underside of a mature earthworm showing the rows of setae.

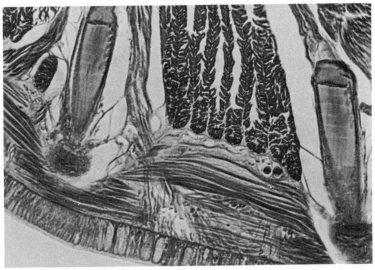

Illus. 30. Cross-section through a pair of setae showing muscles for retraction.

used to anchor the back end of the worm in its burrow while the forward end seeks food or a mate (Illus. 28–30).

Illus. 31. The central nerve pathway with septal branches off the main trunk to the sides.

Encased within the central tube or coelom, is the fifth tube, which is the digestive tract. Stretching from the mouth to the anus, it consists of a number of special chambers for extracting food and digesting it. This food is then turned into energy and into the chemicals needed to keep the cells of the earthworm alive and growing. This fifth tube is really a most important one. It is able to move independently of the other tubes because it is suspended in the fluid sheath

35

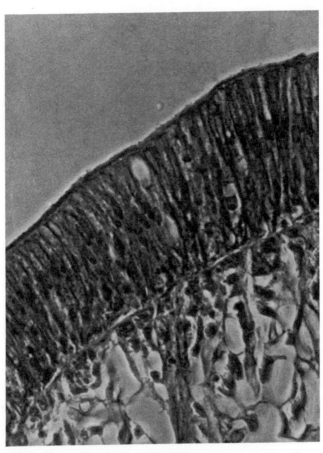

Illus. 32. A much enlarged view of a small segment of the outer muscle layer. One of the sensory cells connected to the main nerve pathway can be seen.

of the coelom. Most of the remaining organs of the earthworm are wrapped around the digestive tube.

Organic material, either animal or vegetable, is sought out by the sense organs around the first segment or ring. This segment looks shovel-shaped from the front. It overlaps the bottom first ring which is actually part of the second segment. The major nerves for sensitivity supply the brain or *ganglion* with stimulations from light-sensitive organs in the outer layer of skin or *epidermis*. There are sense cells for light and sense cells for vibration. They are scattered throughout the outer layer, but most are on the upper surface and near, or on, the anterior end of the worm (Illus. 31 and 32).

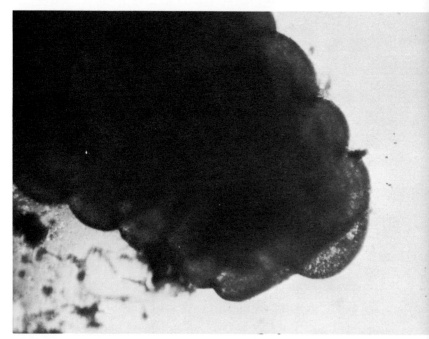

Illus. 33. The first 4 segments of an embryonic earthworm as it moves through the milky-white material within the egg case. The dark shaded area is the beginning of the suprapharyngeal ganglion.

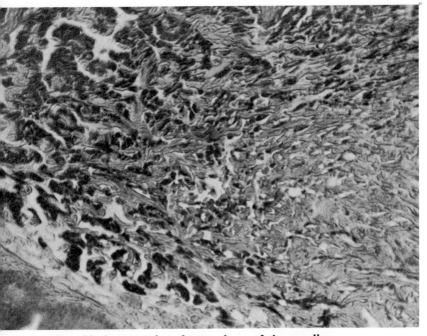

Illus. 34. Cross-section of nerve tissue of the ganglion.

Nervous System

The brain is actually a ring of nerve cells and fibres with a pair of thickenings at the top and a pair at the bottom. Through the middle runs the digestive tract. The top pair of nerve bundles is called the *supra-pharyngeal ganglion* because it is on the top of the pharynx, which is the part of the digestive tube running through the middle (Illus. 33 and 34). The lower pair are called the *subpharyngeal ganglion* because they are beneath the pharynx. The lower or subpharyngeal nerve bundle has two types of side branches. There are

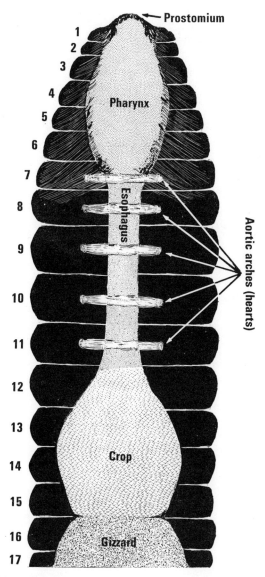

Illus. 35. Digestive system. Numbers indicate rings or segments.

39

septal branches, one set in each segment of the worm from the fourth segment back, and *interseptal* branches. These control the movement of the muscles. Studies have been made to see how fast a nerve impulse travels down the main nerve pathway. It has been estimated that they travel about 1.5 m per second (almost 5 ft per second). This is extremely slow and may account for the slow movement of the segments of worms. The suprapharyngeal ganglion is the actual *brain*, since it is a junction or collective point for all of the nerve pathways, and controls the simple reflexes and movements of the earthworm.

The wave action, the alternate squeezing and stretching, of the segments is timed at about 2.5 cm (1 inch) per second. Although this seems slow, it is actually very rapid, as it is about the speed at which an earthworm passes through soil.

Digestive System

The digestive system is divided into 6 separate organs. The first is the mouth or *prostomium*. This is the muscular organ at the anterior end of the earthworm and is covered on the top by the first segment (Illus. 36). In the fourth segment it passes into the muscle-ringed pharynx. The sucking action of the pharynx draws in the food through the prostomium. From the pharynx, the food-bearing material is passed on to the *esophagus*. This tubular section of the digestive system

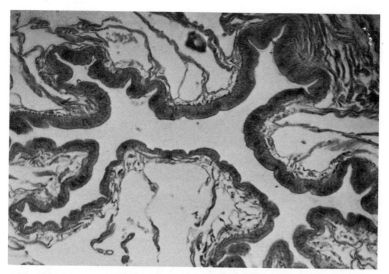

Illus. 36. A cross-section through the 2nd segment, showing the heavily folded and muscular mouth opening used for breaking up organic materials for food.

starts in the seventh segment and runs through the 14th segment. It is surrounded by calciferous (calcium-bearing) glands which filter out certain calcium salts from the food and expel them into the esophagus.

The atmosphere or air in the soil is different from the atmosphere above ground. The worm's calciferous glands aid the earthworm in breathing. The air underground contains very high levels of carbon dioxide gas (CO_2) which is poisonous to most animals. However, the calcium compounds extracted from the soil and formed by the calciferous glands in effect neutralize the harmful effect of the earthworm's breathing air

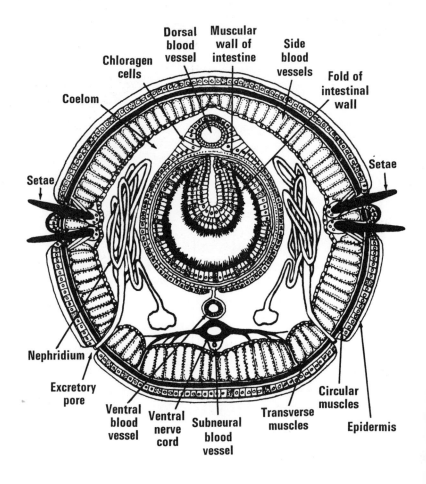

Illus. 37. Schematic cross-section of the intestinal region.

42

Illus. 38. Dissection showing some of the circulatory system and major organs of the digestive system.

containing carbon dioxide. The calcium from the calciferous glands may also combine with the carbonic acid in the blood of the earthworm to form calcium carbonate ($CaCO_3$) and pass out of the digestive system as harmless chalk.

Immediately behind the esophagus and starting in the 14th or 15th segment is a large thin-walled sac shaped like a bottle. This is called the *crop* and it serves the same purpose as the crop in birds. It seems to be a

Illus. 39. Muscles of the gizzard seen from the side.

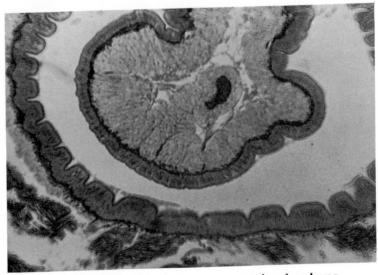

Illus. 40. Cross-section through the strong muscular gizzard area.

storage space for large amounts of food-bearing material which will be pushed backward into the gizzard.

The gizzard is a small tube surrounded by very heavy walls of muscle (Illus. 39 and 40). The earthworm retains sand grains and microscopic stone chips in the gizzard at all times. In this powerful mill, the food of the earthworm is ground to a fine paste-like consistency. The food is made ready for the intestine which begins in the 21st segment and continues for the rest of the length of the earthworm. The last few segments form an organ for excreting the waste products of digestion. This is a very primitive anus.

In the intestine the food is dissolved by chemicals

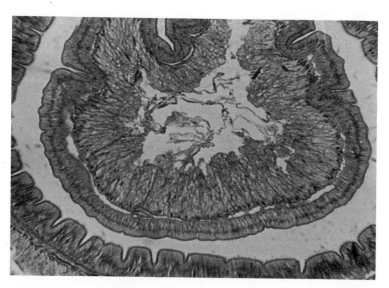

Illus. 41. Cross-section through the intestinal region, showing folds which provide increased surface area.

formed by special cells in the lining. The cells absorb the food directly and transfer it to the blood. The walls of the intestine roll into folds to present the greatest possible surface area for the absorption of food into the blood (Illus. 41). This fold or *invagination* is called the *typhlosole*.

The intestine has other types of loose cells which break down certain chemical compounds in the food. These have not been studied by anyone in detail, but they help the earthworm to manufacture all its needs without the use of sunlight, which most other animals need for vitamin production.

Excretion

The waste products of digestion contain not only a good deal of solid matter which is indigestible, but also some poisons. The most important are the chemical compounds containing nitrogen (N). Although this chemical is very valuable to life and is one of the basic components of all proteins, when the proteins and other substances are digested or broken down by the body, nitrogen poisons are released. Among these are ammonia and nitrous acid.

The earthworm has small tube-like separator organs which take out the nitrogen and its compounds and pass them out of the body through little pores or openings leading to the outside. The pores, which are twisted tubes called *nephridia*, are found in every seg-

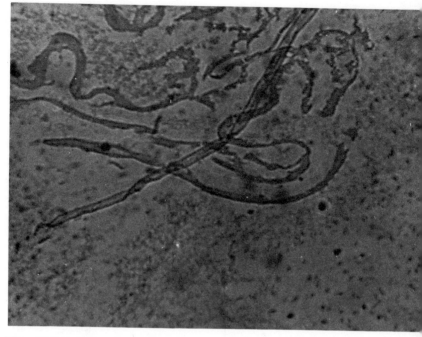

Illus. 42. The thin, twisted tubes of the nephridia.

ment from the fourth to the next to last, or about the
113th in the average earthworm (Illus. 42 and 43). The
nephridia have tubes which lead out to the surface on
the underside of each side of the earthworm. These
tubes are always in the next segment after the one in
which the nephridium is actually located. These
nephridia excrete a constant tiny flow of weak urine
out of the worm's body and into the soil. This aids the
soil and provides chemicals for plants in their growth.

47

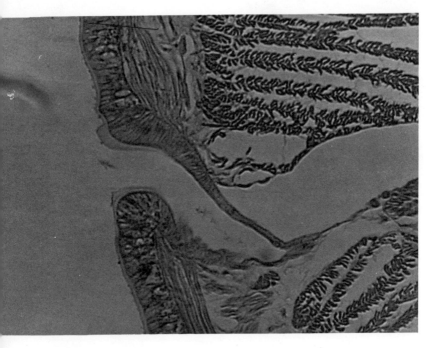

Illus. 43. Highly enlarged photomicrograph showing the nephridium through which nitrogen wastes are drained from the earthworm's body.

Circulatory System

The circulatory system of the earthworm surrounds the esophagus. The circulation of the blood in earthworms serves the same purposes that it does in almost all other animals. The blood itself contains an oxygen-attracting component, *hemoglobin*, which gives blood its characteristic red color on contact with air. Within the earthworm's circulatory system, the blood is normally blue (as it is in human beings also!).

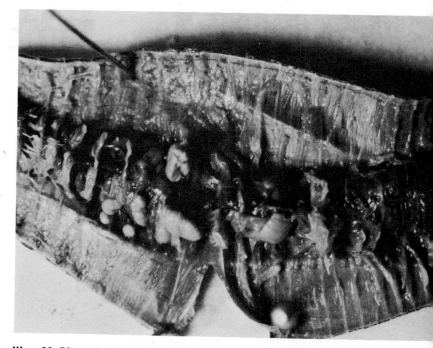

Illus. 44. Dissection from the midline of the 4th through the 30th segment.

There are two main blood-carrying vessels. The distributing vessel carries blood to the muscles and other cells of the segments. Called the ventral vessel, it hangs below the digestive tube just about in the middle of the coelom. The collecting vessel, or median dorsal vessel, lies above the digestive tube and, like the ventral vessel, runs the length of the worm. Another minor vessel suspended below the ventral is known as the subneural vessel.

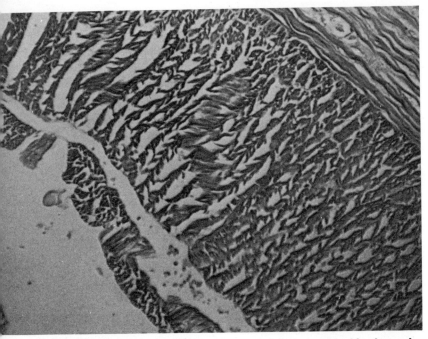

Illus. 45. The cell structure of one aortic arch (heart) and its blood vessels.

The blood is pumped through the system by 5 thick-walled, circular transverse vessels, or aortic arches (Illus. 45 and 46). Each of these is actually a heart. In each segment there are large vessels which connect the dorsal and ventral vessels. There are beds of very fine vessels or capillaries through all of the major muscles and organs so that the blood may supply oxygen and foodstuffs to the cells and carry away carbon dioxide and nitrogen-bearing wastes.

The subneural vessel carries nourishment and oxygen to the main nerve pathway, which runs close to it along

Illus. 46. Close-up of the dissection of the aortic arches (hearts), septa and seminal vesicles.

the length of the earthworm. The hearts beat in systematic fashion and pump the blood through the ventral vessel. The blood travels the circuit of the body and back through the hearts. The earthworm absorbs oxygen directly through its skin into the capillaries below. It is the perfusion of red oxygenated blood through the skin and muscles just beneath the skin which gives the earthworm its reddish tint. For this reason some people commonly call it the bloodworm, although the name bloodworm properly applies to the larval stage of a flying insect. The earthworm must keep its skin moist to continue to absorb gas from the air. It can form mucus on the epidermis or even use some of the coelomic fluid in periods of exposure to drought or light.

The coelom also provides other special capacities to the earthworm's metabolism. The cells at the surface of the digestive tube and around the major blood vessels are modified into special waste receptors called *chloragen cells* (Illus. 47). They absorb waste from the blood, particularly metal-bearing compounds. In time, the chloragen cells break off and float in the coelomic fluid. Some are filtered out by the nephridia, while others are destroyed in another way—they are captured by the *ameboid cells*.

Ameboid cells are free-floating cells with flexible cell walls that can engulf intruders inside the earthworm body. They attack and destroy bacteria and other

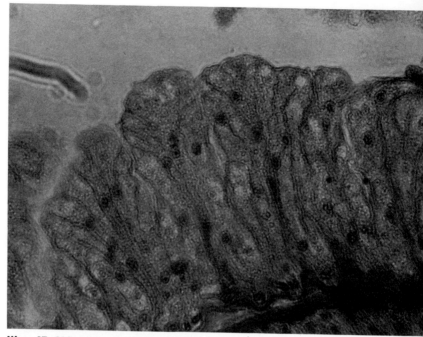

Illus. 47. Chloragen cells still attached to the lining of the digestive tract.

invaders, and also destroy chloragen cells which have become detached after filling up with waste. Metallic waste material is disintegrated and deposited in the epidermis, where it contributes to the color of the worm. This is the reason that newly hatched worms are pink, the blood showing right through the skin, and older worms are dark red and even bluish. Generally, the older the worm the thicker and the darker its segments are.

The digestion absorbs already rotted material. The earthworm cannot break down plant substances directly

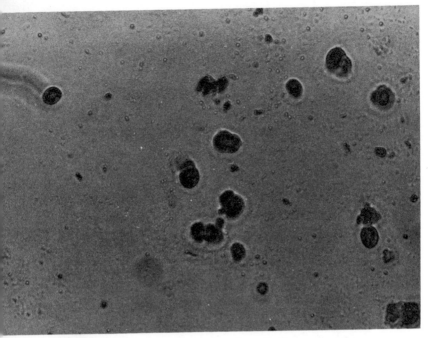

Illus. 48. Bacteria and fungus cells from earthworm castings.

but must rely on bacterial action to prepare its food. In this way, the earthworm is really an intermediate reducer in the ecology system. The digestive system of the earthworm is filled with bacteria which are really still "outside" the earthworm's body, since they are not absorbed. There are nearly 5 times as many bacteria in the earthworm's manure or cast as in the ground around it. In one study done in Germany, it was shown that there were 11,000,000 bacteria per gram of soil and 50,000,000 bacteria per gram of earthworm casting (Illus. 48).

THE EARTHWORM IN ECOLOGY

The earthworm is one of the most important animals on earth. It is the one organism that turns, aerates, builds and fertilizes the soil. Each of these steps is absolutely necessary for the continued growth of new plant life. Since plants are the only manufacturers of food in the form of sugars and starches from the chemicals and sunlight available on earth, they are the basic source of life for all other creatures. The earth-worm is one of the very few organisms that aids the growth of plants with anything else besides its waste products.

Aeration of the Soil

The earthworm does three things of great value to the soil. One of these activities is mechanical and the other two are chemical. First, and most obvious, the earthworm ploughs and turns the earth. This brings to the surface tons and tons of well decayed organic material, at the same time taking underground new organic material. The holes and burrows made by the worm provide passages for air to reach under the soil. Without air becoming dissolved in the soil, wet areas would produce continuous fermentation. This

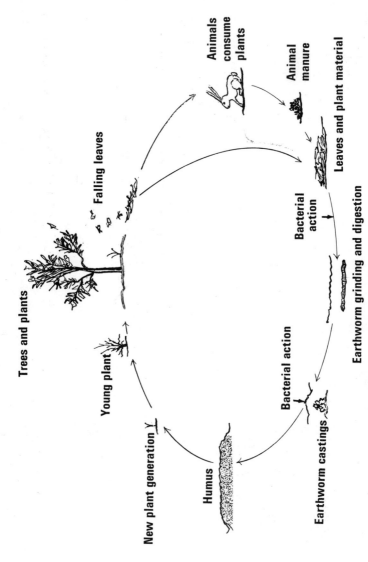

Trees and plants

Falling leaves

Animals consume plants

Animal manure

Leaves and plant material

Bacterial action

Earthworm grinding and digestion

Young plant

Bacterial action

Earthworm castings

New plant generation

Humus

Illus. 49. Humification cycle—from old plant material to new plants.

means that the organic material would be used by bacteria that cannot stand free oxygen and the result would be sour and foul-smelling soil where little or nothing could grow. This often happens in parts of swamps and swampy ground.

Humification

The second action of the earthworm is to grind up and digest organic materials. This reduces dead plant and animal material to the point where its chemical composition can be used by new plants and they can be eaten and used by new animals. Actually, the earthworm is a principal agent in the process called *humification* (Illus. 49). Humus is the mass of dead and partly decayed plant material found on the ground in woods or thickets. There is a much thinner form of humus in pasture lands and meadows (Illus. 50).

If we take the example of a single leaf from, let us say, a walnut tree, we can see exactly the earthworm's role in humification. As summer turns to autumn, the tree begins to slow down its supply of nutrients to the leaves and certain chemical substances called *enzymes* begin to cut off the flow of nutrients from the leaf. The leaf begins to turn color and is attacked by insects.

Insects chew holes called *fenestrations* in the leaf and it soon falls off one windy day and lands on the ground. With rainwater and more leaves falling on top, the leaf becomes covered in a dark and wet environment.

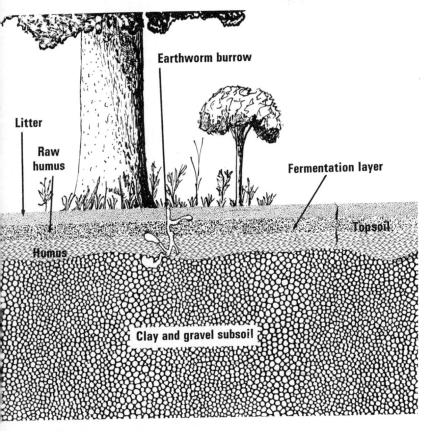

Litter

Raw humus

Earthworm burrow

Fermentation layer

Topsoil

Humus

Clay and gravel subsoil

Illus. 50. Soil layers.

Bacteria begin to break down the woody material into less complex substances and finally into basic organic compounds. By now, the leaf is very soft and dark and beginning to break apart. The earthworm comes up from its burrow under the stack of leaves, or climbs on top of the stack at night, and breaks off small pieces, which it pulls underground (Illus. 51 and 52).

58

Illus. 51. A night photograph catches an earthworm out hunting food away from its burrow.

Illus. 52. An earthworm burrow about 6 inches below the soil level lined with white sandy material which had been put on the surface to show how earthworms take new materials down to lower levels of the soil. This is one day after the material had been placed around the entrance to the burrow.

Illus. 53. A pile of well-rotted humus mixed with castings all thrown out on the surface of the soil by a number of earthworms inhabiting a small area under a tree root.

In the burrows the tiny insects and bacteria of the soil continue to break down or reduce the leaf material still further. The earthworm eats the rotted leaf parts and they are ground to a fine paste in its gizzard and digested by the action of the internal bacteria and fluids in its digestive tube. Finally, the broken-down and nearly reduced organic material from the leaf is expelled in the earthworm's manure (cast) on the surface of the soil (Illus. 53). There is further decay caused by the action of still more bacteria until the chemical compounds are ready to be taken into the roots of a new generation of plants. This last product of reduction is the dark brown, fertile material which mixes with sand and clay to become soil.

There are two definite factors that influence the speed of this humification. One is the proportion of carbon to nitrogen in the plant materials. The more carbon there is in relation to nitrogen, the longer it takes for decomposition by the reducer organisms. These proportions can be stated in numbers with the first number being the amount of carbon, and the second the amount of nitrogen which is always set at 1. For pines the number is 66 to 1, for maples it is 52 to 1 and for elms it is 28 to 1.* This simply means that pine wood and needles take longer to reduce than maple wood and leaves and that the elm is the fastest to decompose of the three.

The second factor that affects the rate at which decomposition or humification takes place is the amount of moisture. The sides and surfaces of all burrows and holes in the soil are covered with a thin layer of water. This is what makes most soil sticky and damp to the touch. Without this layer of sufficient water, none of

* Much of this section is dependent upon W. Kuhnelt, *Bodenbiologie*, Vienna (1950) and F. Schaller, *Soil Animals*, Ann Arbor, Mich. (1968).

the bacteria or other reducer organisms could do their work.

The earthworm has the effect of a mechanical grinder and digester, which speeds up the reduction process. Of special importance is the quantity of *lignin*, the substance which stiffens and strengthens wood. The earthworm first grinds away at the lignin and the bacteria and other microscopic reducers can then digest it.

Fertilization

The third action of the earthworm to fertilize the soil is to excrete nearly reduced organic materials and millions of bacteria upon the surface of the soil (Illus. 54). This adds mixed organic and mineral material to the top layer of the soil, and greatly aids the development of seeds and new young plants. The fertility of soil is directly affected by a lack of earthworms. If you find a grassy patch containing many earthworms, you know that it is good fertile soil and that it will support plant life for many years to come.

The earthworm contributes in these three ways to the building as well as the maintenance of the soil and to the growth of plants. Since it is one of the most ecologically important creatures on earth, it should be protected and aided in its work. One way to protect it is to be very careful about using insecticides and poisons which stay active in the soil. Any insecticide or fungicide (chemical used to kill unwanted insects or

Illus. 54. This calm and still forest floor with rocks and mosses at the base of a big tree is the cover over a seething mass of tiny organisms carrying on the process of humification upon which the tree and mosses depend.

fungi) should be of the type that decomposes in the soil. These are called *biodegradable*, which means that they will be decomposed by bacteria into harmless substances.

The earthworm's habit of digging up the soil has been measured very carefully in many parts of the world. It has been estimated that earthworms add an additional 8 to 15 tons of topsoil per acre each year. This means that they build up the level of the ground at about 17 cm (7 inches) each century. This does not seem like very much, but it is if you consider that in 1,000 year's time earthworms can bury walls and monuments as tall as an adult person. The earthworm is the major reason that so many Greek and Roman remains in Europe are now buried underground.

KEEPING AND WATCHING EARTHWORMS

Earthworms can be kept in small containers for up to a month. They are very easy to feed and interesting to watch. All that is required is a container at least 25 cm (10 inches) deep and filled with moist soil mixed with old well rotted leaves and plant parts, but free of chemical fertilizers or insecticides. The only other requirements are a flashlight and an old spoon for digging and transporting earthworms.

The soil should be pressed down very hard and packed tightly into all corners of the container. Add the earthworms next and put a leaf mulch on top. The container can be put in a cool dark place and the earthworms can be watched for brief periods with the flashlight at night (Illus. 55). After a week, careful digging will turn up a burrow and possibly some egg cases. If the egg cases are buried again and left to stay undisturbed for a month or two they will usually hatch out and the hatchlings can be seen joining their parents to hunt for food on the surface. By adding a piece of old wood or a flat stone on the surface, it will be possible to get the worms to open their burrows underneath it.

Never cut an earthworm or pull off segments, since

Illus. 55. Two earthworms caught mating at night on the ground surface.

it hurts the worm and may cause it to die from the wound. It is difficult to see very much by just cutting up an earthworm—for proper dissection, special chemically treated earthworms must be bought from a biological supply company. It is better to learn from the living animals and watch their life cycle.

If the earthworm container is to be kept inside a house or apartment, make sure there is enough space— at least a hand-width—between the top of the soil and the lip of the container. This will prevent the earthworms from escaping into the room. Earthworms should be kept in captivity only in the spring and summer. They should be returned to the outdoors late in the summer so that they can feed and dig deep for the winter.

EXPERIMENTS WITH EARTHWORMS

A school class, or anyone who is interested, can do some very instructive experiments with earthworms. The three experiments described here can be carried out with a little work and some scrap materials. Experiment 1 is an experimental count. Experiment 2 is raising earthworms in a humus pit. Experiment 3 is constructing and using an observation frame.

1. Experimental count

Lay out a square metre of ground by tying string to 4 corner posts about one third of a metre (13 inches) above ground surface from which all grass and other plants have been cleared (Illus. 56). Quickly but carefully dig down at the line all the way around and put all of the soil dug up into a large container with plastic or other lining so that no earthworms escape. Dig down exactly one third of a metre (13 inches) (Illus. 57), or if you have the time and enough help a whole metre. This will give you either one-third cubic metre or one whole cubic metre. Go through the excavated earth very carefully by hand and separate out all of the earthworms.

In old pasture land or good fertile garden, such as

Illus. 56. The area where the experimental count is to be made is marked out with string to show the 1-metre-square area.

Illus. 57. The area of the experimental count has been dug out to a depth of one third of a metre (13 inches). The soil will be carefully sifted to capture all of the earthworms.

Illus. 58. A mass of earthworms recovered from the experimental area.

the area shown in the photographs, you should be able to find 440 or more earthworms per cubic metre (Illus. 58). In ploughed ground you should find about 290 worms per cubic metre and in new topsoil you will get fewer than 100. Along with earthworms of all sizes you should also find earthworm egg cases. Keep a count of each area and compare them to see which soil conditions are the best for earthworms and thus for humification.

Illus. 59. An earthworm pit two thirds of a metre (26 inches) has been dug and two logs two metres (6 feet 6 inches) in length have been set in each wall of the hole to hold back the soil.

Illus. 60. Earth is packed against the edges of the pit next to the logs.

2. Raising earthworms in a humus pit

Commercial earthworm farms raise earthworms by the millions for sale to farmers and fishermen. You can build a humus pit for raising earthworms in about 5 square metres (6 square yards). First select a spot with good drainage and with tree cover over it, so that the summer sun cannot bake the soil hard. Dig down about two thirds of a metre (26 inches) and set logs, pieces of tree trunk, old railway ties (sleepers) or any other solid materials into the sides of the hole so that the soil will not roll back into it (Illus. 59).

When these are set, back-fill with some of the soil

you have dug out, so that there is a tight wall of earth all around the logs or other material (Illus. 60). Then spread plastic or canvas sheeting all over the bottom (Illus. 61). Pack it tightly into the corners. Punch some big drain holes at the bottom and around the sides and make sure some holes go off into the ground so that these holes in the plastic or canvas can drain off excess water.

Tuck the excess material around the logs and cut it

Illus. 61. A plastic sheet is packed into the hole tightly on all sides and the bottom is punched with drain holes.

Illus. 62. The excess plastic has been tucked under the logs or cut away and raw humus and soil are being mixed in the pit.

off (Illus. 62). Now scatter finely ground or shredded plant material and some rich dark humus across the bottom. Then build up layers of soil and dead plant material until you come to the top layer of soil. Put as many earthworms as you can find in a hole in the middle and rake and water the pit weekly (Illus. 63). It should begin to show the signs of earthworm activity within a week or two and prove to be the most fertile place for them to reproduce.

Illus. 63. The fill is being raked and watered and the earthworm colony being added.

3. An observation frame

An observation frame is a terrarium built especially for keeping and observing soil-dwelling species. The best arrangement for earthworms is to keep the layer of soil as thin as practical. While single-weight window glass gives the best visibility, it also breaks more easily. For safety's sake, it is better to use a single thickness of clear plastic sheeting. To build the observation frame you will need the following supplies:

2 sheets of single-thickness clear plastic 25 cm × 38 cm (10 × 14 inches).

2 strips 25 cm long and 2 cm wide (10 inches × $\frac{3}{4}$ inch).

1 strip 38 cm long and 2 cm wide (14 inches × $\frac{3}{4}$ inch).

1 tube of silicon rubber glue.

CAUTION: Do not get silicon rubber in your eyes. CHILDREN should use it only in the presence of an adult.

Using silicon glue, cement the long (38-cm) strip to the bottom of the longest side of the clear plastic side. Cement the shorter strips on each side to form a sort of three-sided tray. When these three strips are firmly bonded, cover all of the three standing sides of the thin strips with the silicon glue and put on the other large sheet. They should be propped up into an upright position with the opening at the top. After 24 hours the inside should be washed out. A small leak will usually appear, which is desirable for drainage. If one does not appear, drill a small hole up from the bottom.

Put a natural layering of soils into the cured and cleaned observation frame. That is, place clay soil at the bottom, humus in the middle, over the greatest depth, and white sand or fine gravel for a thin coating at the top (Illus. 64). Make a few holes into the sand through the humus layer below. Place one or two large mature earthworms inside the frame and sprinkle with

Illus. 64. The observation frame is filled with clay soil at the bottom, humus in the middle level and white sand at the surface.

the least amount of water. The soil should be packed in fairly hard, to eliminate air bubbles. The whole observation frame must then be covered with heavy brown paper or aluminium foil to make it light-tight. It should be covered completely and put into a cool and dark place. The paper may be removed once a week for observation. Feed your earthworms a few dead leaves and they will surprise you as they dig and plough through the soil (Illus. 65). Perhaps one will even leave an egg case which you can observe until it hatches. The whole frame must be completely washed out once a month and new soil installed in it. It is the best and simplest way to watch the underground activity of earthworms.

Illus. 65. The surface sand has been carried down into the earthworm's burrows and castings are being excreted over the sand left on the surface.

Illus. 66. The last segments of an earthworm are fast being pulled into a burrow as daylight dawns.

INDEX